A Day in Outer Space

Created by James Helmering and Rodney Pritchard

ISBN 978-0-692-36484-0
9 780692 364840 >

DRAWNBYME

Presents

A Day in Outer Space

by James Helmering and Rodney Pritchard

Cover art designed and created by James Helmering and Rodney Pritchard.

Drawn by:

▷ ______________________________

DRAWNBYME

Storybooks where you become the illustrator.

Instructions

Have you ever wanted your name in a published book? Now it's a possibility. This book brings out your inner creativity and allows you to be the illustrator. Written by James Helmering and Rodney Pritchard, drawn by you.

"Look everyone! I'm in a book!"

Step 1 - Use a permanent marker to write your name on the cover of this book next to the "Drawn By Me" logo

Example:

Use a black ink pen to print your name on the inside line (Page 3) under the area that says, "Drawn by."

Step 2 - Read the book and draw whatever you think belongs in each section. There are even suggestions at the bottom of each page that help you think of possible images to draw.

Step 3 - Don't forget this is your book! Let your imagination run free and, as always, have fun.

Looking out my window, I stare at the moon,
and wonder what it does for fun.

Draw the moon looking through a window.

Does it just sit there all day, with nothing to say,
reflecting light off of the sun?

Draw a moon half lit up by the sun.

There must be a way for me to play up there with all the stars.
If I leave right now, I could be back before bed.
Surely it can't be too far!

Draw a picture of the stars.

I pretend to fly, straight up to the sky, leaving the earth behind.
I'll spend just one day in the Milky Way. I can't wait to see what I'll find.

Draw a spaceship entering space.

My ship enters space and my heart starts to race,
as I see all these marvelous things.
The Sun is the first one to catch my eye
and then I see Saturn's rings.

Draw a picture of the sun.

Mercury it seems is the best place to start, since it's the first in line.
To see all the planets, I'll have to hurry. I don't want to waste any time.

Draw a picture of the closest planet to the sun, Mercury.

I do a loop around Mercury and then off to Venus,
because it's the second to see.
Then I pass Earth which is third from the sun
and is home to you and me.

Draw a picture of Venus or Earth.

As I pass Mars, I gaze at the stars and my heart starts to melt.
I'm so lucky to have seen these amazing things;
up next is the asteroid belt.

Draw a picture of Mars or the Asteroid Belt.

Jupiter is the largest planet I've seen; at least in the Milky Way.
I'd fly around it just once if I could,
but that would probably take me all day.

Draw a picture of Jupiter.

With five planets down and three to go, the low fuel warning dings.
Saturn is the sixth planet from the sun and has many colorful rings.

Draw a picture of Saturn.

I'm almost to Uranus; it's hard to explain,
all the fun I've had on this ride.
Did you know that Uranus is special
because it actually spins on its side?

Draw a picture of Uranus.

I bet mom's getting worried, about where I've been,
I should start to head home soon.
I give the boosters a flip and make a quick trip
around the last planet, Neptune.

Draw a picture of Neptune.

I make my way back past all of the planets; I really wish I could stay.
I'll have to return someday soon, when I have more time to play.

Draw a picture of you flying home in your rocket ship.

When my rocket ship lands, I wave my hands
letting everyone know that I'm back.
I gather my things hop off my ship,
now it's time to go and unpack.

Draw a picture of your rocket ship on the ground.

I'm happy and safe on Earth's ground,
and lucky my coordinates landed me back in my town.

Draw a picture of the town or city you live in.

I hear mom yell for dinner, so I run to the kitchen,
I can't wait to see her face.
Just wait till I tell her that I just got back from a day in outer space.

Draw a picture of your favorite thing in outer space.

The End

Calling All Young Artists!

We need you to share your artistic gift with the world.
Collect all of the Drawn By Me books and showcase your work to friends and family.

www.DrawnByMeBooks.com

Storybooks where you become the illustrator.

A Day at The Beach
A DAY AT THE FARM
DRAWNBYME
A Day at Grandma's
A Day at The North Pole
Written by
DRAWNBYME
A DAY AT PLYMOUTH ROCK
DRAWNBYME
A DAY IN THE WILD WEST
WANTED
DRAWNBYME
A Day At The ZOO
Written by:
DRAWNBYME

Created by James Helmering and Rodney Pritchard

ISBN 978-0-692-36484-0

Made in United States
North Haven, CT
05 November 2026

21765613R00017